# Common Confusions In Macroeconomics

## Peter Gutmann

Bloomington, IN  Milton Keynes, UK

authorHOUSE

*AuthorHouse™*
*1663 Liberty Drive, Suite 200*
*Bloomington, IN 47403*
*www.authorhouse.com*
*Phone: 1-800-839-8640*

*AuthorHouse™ UK Ltd.*
*500 Avebury Boulevard*
*Central Milton Keynes, MK9 2BE*
*www.authorhouse.co.uk*
*Phone: 08001974150*

*First published by AuthorHouse 3/17/2006*

*ISBN: 1-4259-1544-2 (sc)*

*Library of Congress Control Number: 2006900492*

*Printed in the United States of America*
*Bloomington, Indiana*

*This book is printed on acid-free paper.*

# PREFACE TO

# "COMMON CONFUSIONS IN MACROECONOMICS"

This book clarifies common confusions in macroeconomics. It does not use equations, graphs, diagrams or footnotes.

The book is designed to focus on a number of macroeconomic subjects that are so often unclear in public discussion of policy, in the press, and in economics textbooks. The book also presents information on US income distribution, as well as historical data on inflation rates, on real GDP per capita growth rates and on population growth rates.

It covers a series of important topics. Included are : "surplus of savings"; effects of the import surplus; steep and shallow yield curves; capital movements and interest rates; overvaluation of the dollar; deficits and debt; world income redistribution and petroleum prices; the decline in assessment of risk; bubbles; the "twin deficits"; the Achilles heel of the US economy; and more.

New York, January 2006

# TABLE OF CONTENTS

PREFACE ...................................................................v

1.  THE COUNTRY AND THE WORLD ............................1

2.  SMALL GOVERNMENT AND BIG GOVERNMENT,
    BALANCED BUDGETS AND DEFICITS........................5

3.  INCREASE IN DESIRE TO SAVE................................9

4.  "SURPLUS OF SAVINGS" ....................................... 11

5.  THE INTEREST RATE, MONEY AND SAVINGS.......... 15

6.  HOUSEHOLD SAVINGS AND BUSINESS SAVINGS... 17

7.  EFFECTS OF LIQUIDITY ON CONSUMPTION AND
    INVESTMENT ............................................................. 19

8.  PETROLEUM PRICE INCREASES, WORLD INCOME
    REDISTRIBUTION AND WORLD SAVINGS.................21

9.  FULL EMPLOYMENT NATIONAL OUTPUT AND
    CAPACITY NATIONAL OUTPUT .................................23

10. PREDICTIONS: TRENDS AND BLIPS ........................25

11. OF DEFICITS AND DEBT...........................................27

12. THE FED, INTEREST RATES AND SIGNALS..............29

13. NOMINAL AND REAL INTEREST RATES ...................31

14. INFLATION INDEXES.................................................33

15. ADJUSTMENT PERIODS: FINANCIAL MARKETS AND
    REAL MARKETS .........................................................35

16. CROWDING OUT .......................................................37

17. STEEP VS. SHALLOW YIELD CURVES ....................39

18. THE "TWIN DEFICITS"..............................................43

19. THE EFFECT OF INTEREST RATE DIFFERENTIALS ON CAPITAL MOVEMENTS BETWEEN COUNTRIES.45

20. SMALL COUNTRIES AND LARGE COUNTRIES.........49

21. OVERVALUATION OF THE DOLLAR...........................51

22. TYPES OF EXCHANGE RATES..................................53

23. FIXED VS. FLOATING EXCHANGE RATES.................55

24. DIFFERENT GROWTH RATES IN DIFFERENT COUNTRIES................................................................57

25. CONVERGENCE ......................................................61

26. ECONOMIC GROWTH, PER CAPITA GROWTH AND GROWTH IN LEISURE..............................................65

27. LABOR PRODUCTIVITY GROWTH............................67

28. CAPITAL GAINS, INFLATION AND TAXES .................69

29. BUBBLES ................................................................71

30. INCOME DISTRIBUTION ..........................................73

31. WHO PAYS FEDERAL TAXES IN THE US..................77

32. SOME FACTS ON US INFLATION IN THE "COST OF LIVING" ...................................................................79

33. SOME FACTS ON US PER CAPITA GROWTH RATES IN REAL GDP ............................................................81

34. SOME FACTS ON US POPULATION GROWTH RATES. 83

35. BEHAVIOR AND THEORY ...........................................85

36  POLICY: THE SHORT RUN AND THE LONG RUN .....87

37. THE ACHILLES HEEL OF THE US ECONOMY ..........91

APPENDIX.............................................................................97

# 1.

# THE COUNTRY AND THE WORLD

There is a well known relationship that applies to any country:

Private Sector Savings equal Private Sector Investment plus the Government Deficit plus the Export Surplus (or minus the Import Surplus)

This relationship can also be stated as:

National Savings (Private Sector Savings minus Government Sector Dissavings) minus Private Sector Investment equals the Export Surplus

or

National Savings plus the Import Surplus minus Private Sector Investment equals zero

When a country collects its statistics after the end of the year, these relationships (which are identical) hold exactly, since they are actually identities. However, since statistics are imperfect, there will have to be an "errors and omissions" account to take care of discrepancies due to measurement error.

For the world as a whole, in principle, total exports equal total imports. But again, statistics are imperfect, so the actual figures must be adjusted with an "errors and omissions" account. (Actually, due to widespread underinvoicing of exports, total world measured imports will exceed total world measured exports.)

So, for the world as a whole, since exports equal imports when properly measured, the export and import numbers for the individual countries, when summed, cancel out. The world equation is as follows:

> Private Sector Savings in the World equal Private Sector Investment in the World plus the World's Net Government Deficits (sum of the Government Deficits in the World minus sum of Government Surpluses in the World)

Total private sector investment in the world is determined by the business outlook and by financing costs in the different countries. Total net government deficits in the world are determined by a variety of economic and political conditions in the world's countries

So, we conclude that world private sector savings depend on the sum of the world's private sector investment and the world's net government deficits.

As a result, when the sum of the world's net government deficits and the world's private sector investment rise from one year to the next, the total of the world's private sector savings will also rise.

One small confusing element has to do with the treatment of investment carried out in nationalized industries Often these are lumped in with other government expenditures in calculating the government deficit (or surplus). However, if the nationalized industries are not included in calculating the government deficit (or surplus), then the investment projects of such industries have to be included in private sector investments.

It is also possible to establish a world concept akin to the "national savings" concept. Then,

World savings ( The World's Net Private Sector Savings minus the World's Net Government Sector Dissavings) equals the World's Private Sector Investment

This formulation is correct and looks simpler, but is less helpful, since it lumps together the private sector savings and the government sector dissavings. As a result, it is not as clear that private sector savings depend on both private sector investment and public sector deficits.

# 2

# SMALL GOVERNMENT AND BIG GOVERNMENT, BALANCED BUDGETS AND DEFICITS

Many are in favor of big government; others are in favor of small government.

Many are in favor of government that can run deficits; others are in favor of balanced budgets nearly all the time.

These two distinctions are entirely different. Obviously it is possible to have big government with the ability to run deficits (the current situation) and big government without the ability to run deficits. Similarly, it is possible to have small government with or without the ability to run deficits.

Those in favor of big government are generally in favor of government that provides many services and is very active in programs to redistribute income from richer to poorer. Conversely, those in favor of small government want many of these functions to be performed by the private sector, or not be performed at all.

Among the many economic sectors in dispute by big and small government adherents are pensions, medical benefits including drugs, and education. These are all big, expensive and contentious. To finance these, governments need higher and higher taxes. So, those in favor of small government include a high proportion of the higher income groups who have to pay much of the higher taxes to finance big government. As a result, this dispute also becomes a matter of dispute on income distribution.

The second dispute - between those who are willing live with government deficits and those who want balanced budgets - is quite a separate matter. It is ideological in a different sense. For those who want a balanced budget, there is very often an implicit equation between the individual and the state.

The individual usually cannot run deficits, keep on borrowing and run up debt indefinitely. But governments are in a different position. Mostly, they can and do run deficits indefinitely, borrowing more and more, and keep running up debt. They can do this indefinitely, as long as the national product increases over the years, as well as the national debt. As long as the ratio of government debt to GDP does not get out of hand, deficits can continue. In practice, this means that deficits cannot be very "large". But if "large" deficits cause debt to grow faster than income, there comes a point where this cannot continue. This point differs for different countries. Still, beyond a debt ratio of somewhat over 100 per cent, most countries may be forced to change their deficit policies

In the US, proposed constitutional amendments have generally been phrased only in terms of limiting government deficits under most circumstances. As a result, they have not satisfied the small government adherents.

Big vs. small government is a political decision. The facts in the last half century or so are clear. The vast majorities of governments are bigger now than they were prior to the Second World War.

Balanced budgets vs. deficit budgets is an economic decision. Ever since the 1930's it has been clear that there are times in any country, when an unbalanced budget, i.e. a deficit budget, is the best policy.

But, whether it is good policy to have budgets always unbalanced is quite another question. The facts show that in the past half century most governments have run deficits nearly all of the time.

# 3.

# INCREASE IN DESIRE TO SAVE

What happens when there is an increased desire to save by households? This might be due to a rise in uncertainty due to a financial crisis as occurred in some Asian countries in the latter 1990's. Or it might be due to an increased desire to buy private housing units as in China in the 1990's. Or it might be due to the aging of the population facing the costs of retirement such as the situation in Japan in the 90's and in the first decade of the 21st century.

An increased desire to save is the same as a decreased desire to consume. When consumption declines, less is sold, supply of output becomes greater than demand for output, inventories pile up, production decreases and GDP drops.

The end result is lower GDP, lower employment and greater unemployment. The level of GDP will drop until demand for output will again equal supply of output. At this new level, the total amount of saving is very likely to be less than before due to lower income and higher unemployment. This is a well known phenomenon; it is called the paradox of thrift.

Peter Gutmann

A somewhat different scenario has been suggested by some economists. This scenario involves changes in government behavior. But this scenario is not automatic. It depends on specific government choices that do not necessarily occur.

As consumption drops, government may choose to take up the slack by consciously increasing its deficit through increased government expenditure - thus increasing government demand for goods and services - and/or decrease taxes - thus increasing consumer demand. If these events unfold, the increase in the government expenditures and/or decrease in taxes can completely counterbalance, more than counterbalance or less than counterbalance the decrease in consumption caused by greater desire to save. Net results of such government action can be no change in GDP, increase in GDP or decrease in GDP respectively. It should be emphasized that such outcomes depend on conscious government reaction to an increase in the desire to save.

An increase in the desire to save can also be caused by an income redistribution from those who save little or nothing to those who save a substantial amount. In practice, this means an income redistribution from the lower income groups to the upper income groups, say due to taxation policy.

# 4.

# "SURPLUS OF SAVINGS"

The term "world surplus of savings" has become a somewhat popular "explanation" of low long term interest rates. But is this "explanation" valid?

As noted in chapter 1, the following relationship holds on a world total basis:

World Private Sector Savings equal World Investment plus the sum of Net Government Deficits in the World

When the sum of world investment and net government deficits in the world go up, world savings also go up.

We have to distinguish between: (a) increased desire to save and (b) increase in actual savings. These are not the same.

(a) What happens when there is an increased world desire to save, i.e. a decreased world desire to consume? There will then be economic adjustments downward in world income, so that world private sector savings again equal world investment plus the sum of government deficits in the world This well known "paradox of thrift" applies to the world as a whole as well as to individual countries. It says essentially that an increased desire

Peter Gutmann

to save will reduce total world income but will not increase world actual savings.

In practice it is even more complicated. The reduction in world income will increase the deficits of the world's governments, and most likely also reduce world investment (due to deterioration of future outlook).

So, an increased desire to save (i.e. decreased desire to consume) is a negative factor for world income. A "surplus of savings" in the sense of desired savings, as a negative for world income, will also tend to be a negative for interest rates for all maturities in many countries.

However, this is not the sense in which "surplus of savings" is usually used. Usually it is used in the sense of "world surplus of actual savings".

(b) We have seen that actual world savings will be determined by actual world investment plus the sum of net government deficits in the world.

When business outlook becomes more favorable, world investment will go up, but government deficits will tend to go down. Still, the rise in investment will tend to exceed the fall in government deficits (otherwise world income will not go up). As a result, actual world savings will rise.

Similarly, an increase in the net government deficits in the world will increase world savings (provided that any fall in world investment is a lesser amount).

With higher world income, short term interest rates will generally go up in most countries, and so ought long term interest rates, though to a much lesser extent.

It is claimed, as of 2005, that a world "excess of savings" caused long term interest rates to go down instead of up.

How can this be? All of the higher world savings that come with higher world incomes are absorbed by higher world investment (probably compensated in part by lower net government deficits) in the world.

What financial form do higher world private sector savings take? The counterpart of annual private sector world savings are increases in world assets such as government short, medium and long term obligations; corporate bonds, short term obligations and stock; newly constructed houses and apartment buildings, office buildings, factories, etc.

This increase in assets joins the much larger stock of existing assets, The question then becomes this: what is the demand for the various classes of assets, the various maturities of obligations, the various risk categories?

In 2005 it was widely noted , from the chairm of the Fed on down, that financial investors were increasingly ignoring risk. In other words, in the risk/return relationship, the risk of holding long term obligations was widely downplayed. As a result, the demand for long term obligations soared, the price of such obligations rose and the yield declined. This had not been anticipated at the start of the year by the Fed, by brokerage firms and by financial advisors.

This downplaying of risk was particularly true of foreign central banks in 2005, notably those of China, Japan and other Asian countries, and, later that year, the central banks of oil producing countries. Due to their large foreign trade surpluses, they purchased very large amounts of the Treasuries (bills, notes and bonds) of the US government and those of other foreign countries. The outcome was that the price of the long term Treasuries (which yielded the highest interest rates) went up instead of down and the yield went down instead of up, i.e. the yield curve flattened.

So, we can conclude that it was not a "world surplus of savings" but a world "discounting of risk" that caused low long term interest rates.

# 5.

# THE INTEREST RATE, MONEY AND SAVINGS

The interest rate is the price of money per period of time. It is determined by a demand/supply relationship. When demand for money increases, the interest rate goes up. When supply of money increases, the interest rate goes down.

The Fed determines, to a large degree, the US supply of money. For that reason, the Fed strongly affects the short term interest rate. Indeed, it determines the very short term interest rate, the Federal Funds rate, the interest rate banks charge to each other on overnight borrowing.

Savings are an altogether different concept. Private sector savings are composed of household savings and business savings. Household savings equal after-tax household income minus household consumption. Net business savings equal business after-tax profits minus dividends, i.e. retained earnings. Gross business savings equal retained earnings plus depreciation.

A popular theory in the early years of the last century and before held that the interest rate is determined by the demand

and supply of savings. There are still a few who cling to this long abandoned theory. Many studies have shown that the amounts that households and business wish to save are essentially unaffected by the interest rate. There appears to be no direct relationship between the interest rate and savings. So, the interest rate cannot be determined by the demand and supply of savings As noted earlier, the short term interest rate is determined by the demand and supply of money, with the supply side of the equation essentially determined by the Fed.

# 6.

# HOUSEHOLD SAVINGS AND BUSINESS SAVINGS

Textbooks, as well as newspaper articles, continuously analyze and discuss household savings. These have been remarkably low in the US for years, and in some periods actually negative. This is a serious concern.

But, there are other kinds of private sector savings, namely business savings. Total private sector savings is the sum of household and business savings. Business savings, otherwise known as after-tax retained earnings, have not been low. In many years, business savings have provided the bulk of private sector savings. US households, it turns out, are great consumers, but poor savers.

Business savings tend to fluctuate with business profits. The higher are business profits, the higher will be business savings; the lower are business profits, the lower will be business savings.Business profits in turn are largely dependent on national income. This means that business savings will be higher when national income is higher and lower when national income is lower.

Peter Gutmann

In the US, when household savings have been very low, business savings bank very important in "national savings" (private sector savings minus government sector dissavings). And in the US, where domestic investment is greater than "national savings", the shortfall is made up by importing savings from abroad through a large import surplus. So, not only would higher household savings reduce the import surplus, but so would higher business savings. However, in both cases, this risks recession.

# 7.

# EFFECTS OF LIQUIDITY ON CONSUMPTION AND INVESTMENT

When a homeowner borrows funds by remortgaging his house, his net asset position does not change. His wealth remains the same. But his behavior changes.

Similarly, when a business borrows funds from a lender, its net asset position does not change. But its behavior changes.

Behavior of households does not only reflect household wealth; it also depends on the composition of that wealth. When a home is remortgaged with a new mortgage that exceeds the old mortgage, and excess funds are taken out, the asset position of the homeowner becomes more liquid. He has greater debt, but - in effect - part of the value of his house has turned into liquid assets, namely cash. Cash can be spent directly; the equity in a home cannot. But, a home equity loan or a home equity line of credit also creates actual or potential liquid assets, equal to any increase in debt. These liquid assets, i.e.. cash, can also be spent.

This ability, with modern methods of finance, to turn illiquid assets into liquid assets, has become very important in stimulating household consumption. In effect, the home has become a piggy bank. In the first half of the first decade of the 21st century, the increase in household consumption fueled by a remortgaging boom, was vital in pushing up the US economy. Of course, this was also helped tremendously by the very large increase in home values in that era, This increase in housing prices had the double effect of increasing household wealth and thus permitting even greater remortgaging, and rise in household liquidity.

Borrowing on credit cards for consumer purchases is somewhat similar in the sense that it confers potential liquidity which then turns into actual liquidity with simultaneous borrowing. The end result of consumer borrowing, increased liquidity, followed by increased spending on consumer goods, is a decline in net assets (assets minus liabilities) compensated by a higher living style.

Business borrowing by itself does not change net assets either, since rising debt is offset by rising assets. But it does change the liquidity position of the business, hence business behavior.

The business, after borrowing. is in a position to acquire more assets using the new found liquidity. So, assets go up, liquid funds go down, debt remains as is. Net assets have not changed.

This behavioral change caused by the ability to borrow is very important for the economy. It permits business to undertake more investment projects, hence results in higher national investment, an extremely important part of GDP. Liquidity stimulates investment.

# 8.

# PETROLEUM PRICE INCREASES, WORLD INCOME REDISTRIBUTION AND WORLD SAVINGS

Petroleum price increases shift world income from consumers to producers. Exporters of oil are relatively few in number compared to importers. The Middle East, Venezuela, Nigeria, other OPEC members, Russia, and still other exporters of oil have been the beneficiaries of substantial price increases of their exports.

Importers, on the other hand, have had to pay substantially more for oil. Among the importers, the US has been the largest. The result of the price increases of oil has been international redistribution of income from the consuming countries to the producing countries.

What is the economic effect of this international income redistribution? The latest  redistribution took place relatively quickly in the year 2005. The limited number of large oil export countries received a correspondingly large income increase akin to a windfall, but very likely one that would continue more or less in future years instead of disappearing. Increases in world demand by rapidly growing China, India and other countries forced near capacity output in the producing export countries, resulting in somewhat precariously balanced demand and supply, accompanied by the much higher oil prices.

The producing export countries chose not to spend all of the sudden increase in their income. They chose to save a substantial part of it. This policy had been strongly urged by the IMF and other international bodies. So, the shift of world income from the consuming countries that imported oil to the producing countries that exported oil resulted in a net world increase in the desire to save.

An increase in the desire to save part of income is the same as a decrease in the desire to consume. This redistribution of income from those countries that have a lower propensity to save to those countries that have a higher propensity to save reduces world consumption. This, in turn, has a negative impact on world income (see chapter 3). If large enough, it can lead to a world recession.

Of course the negative effect on the world economy lasts only as long as the exporting countries choose to increase their savings substantially. With the passage of time, at least as shown in the 1970's, the increased saving effect will diminish and eventually vanish. The negative effect on the world economy then vanishes as well.

# 9.

# FULL EMPLOYMENT NATIONAL OUTPUT AND CAPACITY NATIONAL OUTPUT

Full employment national output and capacity national output are not identical.

The nation is quite often at an output level where we have "full employment" of labor. But it is very rarely at its capacity output level.

Capacity output is only reached in the course of a long, major war. The last time the US was at more or less capacity output was in late 1944 or early 1945. The other major combatants in World War II such as Britain and Germany also took several years to reach capacty output.

All of the major World War II combatants had absorbed the lessons of the first World War. They realized that, in a market economy, producing at or near capacity output would propagate a great deal of inflation due to a variety of shortages under such circumstances. As a result, they all introduced drastic changes, including rationing, price controls and resource allocations.

In contrast, not only is "full emplyment" output often reached, but economies can and do produce above that level from time to time.

What is "full employment" in a country? Opinions of economists differ for any one period, and also change with changes in the structure of an economy over a period of time. In the US in the mid 'nineties many thought of full emplyment output as the level reached with an unemployment rate, as measured by the household survey, of around 6 per cent plus or minus a fraction. By the middle of the first decade of the 21st century, most economists had modified their view of full employment as around 5 1/4 per cent unemployment plus or minus a fraction.

There does exist a technical definition of "full employment" output. It is that level of output where expectations of future inflation equal the actual inflation subsequently experienced, i. e. where expectations of inflation turn out to be correct. Since US labor markets became more flexible and inflation declined in the decade prior to the midpoint of the first decade of the new century, the unemployment rate corresponding to "full employment" also declined.

# 10.

# PREDICTIONS: TRENDS AND BLIPS

One swallow doesn't make a summer. How about two? Three?

This is a common problem in making economic predictions, to distinguish between a trend and a blip. Economic data series rarely go steadily up , down or sideways. If they go up, they will go up erratically. If they go down or sideways, it will also be erratic.

So, how are predictions made? Does a single uptick indicate a trend? How about upticks in two successive months? Some business economists will make an early prediction before the herd follows, to gain the publicity and prestige that a correct early prediction can yield. But, the earlier the prediction, the greater the chance of error and the negative publicity that it entails. Most business economists prefer safety in numbers. They tend to wait until their peers at other firms are prepared to make the same or similar prediction. That could require three, four or more upticks.

A particular problem arises when an upper turning point is expected sometime in the future. Upper turning points are

Peter Gutmann

notoriously difficult to predict. Such predictions are in error much more frequently. The basic problem is this: how many downward blips make a reversal of trend? As indicated, the record of business economists on predicting upper turning points is poor.

# 11.

# OF DEFICITS AND DEBT

There are several major questions relating to deficits and government debt.

Does government borrowing to finance a deficit have to be repaid at a future time? Does a deficit always increase the burden of the national debt? How high can the debt ratio (debt/ GDP) rise in most countries without serious consequences?

The answer to the first question, "do our children have to repay the government debt we incur now?", is "no". Did we repay the government debt incurred by our fathers? No. Did our fathers repay the government debt incurred by our grandfathers (the World War II debt)? No.

Even the most casual observation of government behavior in the highly developed countries such as the US shows that nearly all countries have deficits nearly all the time. Their debt keeps growing in absolute size. They simply roll it over, i.e. they borrow on new debt to repay the old debt. Very occasionally, there is actually a government surplus. In all the years since the end of World War II, the US has had a Federal Government surplus (hence, debt repayment) in only half a dozen years. All the other years were in deficit with rising Federal Government debt.

That brings us to the second question, "does a deficit increase the burden of the national debt?". To answer this question we have to realize that national GDP is growing, both in nominal and real terms, practically all the time. Obviously, a deficit means that the national debt is growing. But the critical question is this: which is growing faster, the national debt or the country's GDP?

If deficits are "small" and the national debt is growing more slowly than GDP, the ratio of debt to GDP will decline and the burden of the debt will fall. If, on the other hand, the deficit is "large" and the national debt is growing faster than the GDP, the debt/GDP ratio will rise, and the burden of the national debt will rise. It all depends on which is growing faster, GDP or national debt.

How high can the debt/GDP ratio go without causing the country severe difficulties? This depends somewhat on the nature of the country concerned, but debt ratios over 100 percent deserve a close look. When debt ratios ascend above that level, financial investors in that country's obligations are likely to demand higher interest rates at some point to compensate for greater risk. Countries such as Italy and Belgium have had to contend with these issues. But Japan's economy has proven sufficiently different to be able to contend with very high debt ratios without highly negative effects.

# 12.

# THE FED, INTEREST RATES AND SIGNALS

There are still many who believe that the control over interest rates , specifically short term interest rates, by the Fed is the be all and end all of Federal Reserve policy.

But, this is not the case. A quarter point change in the short term interest rate by itself, after all, is likely to have a very limited impact on borrowing costs relatively to the already existing cost of borrowing. And it may have almost no impact on the medium and long term borrowing costs that are important to business and to investment, since the Fed does not control the medium and long term interest rates.

Rather, the Fed sends powerful signals to the economy. These include the Fed's future intentions on interest rate policy as well as the Fed's thinking on current and future developments such as inflation and employment.

The Fed sends these signals in several ways. At the time of the Open Market Committee meetings, the Fed not only announces its new Federal Funds (overnight lending by banks to other banks) interest rate target, but also issues a statement whose language is carefully drawn, indicating its future intentions. For example, in 2004 and 2005, these statements clearly indicated the Fed's intentions to keep on raising the short term interest rate by modest, quarter point, steps at future meetings of the Open Market Committee at the usual 6 week intervals.

In addition, the minutes of these meetings, which are released with a time lag, lay out the discussion among the participants in more detail.

On top of all this, governors of the Fed and Fed regional bank presidents make speeches which also point to future Fed policy. Ocasionally, though, the markets may misinterpret these.

Finally, the Chairman of the Fed testifies before the Committees of Congress at intervals, both with formal statements and responses to questions from Senators and Representatives.

It should also be mentioned that the Fed can impose a variety of direct controls over the banks it supervises. That, too, is available when needed.

All this influences the financial markets and, with that, the economy in more powerful ways than interest rate changes alone would indicate.

# 13.

# NOMINAL AND REAL INTEREST RATES

Nominal interest rates are the rates published in the press. Real interest rates are adjusted for inflation.

There is a well known relationship, the Fisher equation, used to determine real interest rates: the real interest rate equals the nominal interest rate minus the expected rate of inflation (or plus the expected rate of deflation). The real interest rates over different periods of time - short term periods, medium term, long term - have to be calculated for the expected inflation rates over these different time periods.

An example will suffice. If the nominal interest rate for a one year debt instrument is 5% and the expected inflation rate for the year is 3%, then the real interest rate for the year is 2%. So, if $100 is lent, then $105 is received at the end of the year. But $3 of this amount is required to compensate for inflation, so that only $2 is the net real return.

This picture must be adjusted for taxable accounts. If the tax rate is 30 per cent, the net after-tax amount received at the end of the year is only $103.50. This means that the real after-tax interest rate is only 0.5%. Indeed, there have been quite a number of periods when after-tax rates of return on short term interest rates have been less than nothing. This is largely the result of a tax system not adjusted for inflation.

When business makes investments, it is the longer term real rates that are the relevant rates in determining the cost of capital. Hence it is necessary to determine such rates. This is considerably more difficult than determining real short term rates, since the expected longer term rates of inflation are hard to predict. There is a wide variability in estimates by different observers and, in practice, a wide range of error.

# 14.

# INFLATION INDEXES

There appears to be a considerable amount of confusion over the measurement of inflation.The general public has difficulty interpreting the difference between an index of inflation and a "core" index of inflation. In addition, there are a number of different indexes.

The Fed has largely, but by no means entirely, been using a core index of inflation. Such an index strips out of the index energy prices and fuel prices. The logic is simple. Energy prices and food prices tend to fluctuate a lot, both up and down. So, a core index shows more stability than a regular inflation index. This has an advantage in the determination of interest rates. Such determination of interest rate policy by the Fed. should not be impacted by month to month, and quarter to quarter, fluctuations in both directions in the rate of inflation due to food and energy price fluctuations.

However, as the year 2005 proceeded, energy prices went up and up, with fluctuations to be sure, and it became more and more obvious that energy prices were very unlikely to return to the old price levels prior to that year. And so, the rate of inflation indicated in a core index became substantially lower than the rate the public actually experienced. At the same time, the justification of using a core index that failed to recognize the large permanent price rise of energy became less and less. Hence, the public paid less and less attention to a core index of inflation and more and more attention to the (higher) inflation index including food and energy, since it more accurately reflected the position of the consumer.

There are a number of consumer inflation indexes. The favorite index of Alan Greenspan, Chairman of the Fed until expiration of his term at the end of January, 2006, was the core PCE (personal consumption expenditure) index  The actual PCE index itself, of course, does include energy and food. Then there is the well known CPI (consumer price index), in both the actual and core varieties. These indexes may also be available for different geographical areas and for different consumer groups.

Also, we have other inflation indexes. For example, the total GDP deflator inflation index measures the degree of inflation in the GDP totals in the national income statistics. The PPI (producer price index) is another example. It measures the inflation rate of wholesale prices of finished goods.

In general, looking at any inflation index, we should ask: (a) to what group does it apply (particular consumer groups, producers, the whole country, etc.); (b) what is included in the index (all goods bought by consumers, all but fuel and energy, etc.); (c) how should the index be used?

# 15.

# ADJUSTMENT PERIODS: FINANCIAL MARKETS AND REAL MARKETS

When there is a change, markets have to adjust. This adjustment to the new circumstances may be rapid or slow.

Generally, financial markets adjust very rapidly, most almost instantaneously. Real markets adjust slowly.

Many common economic concepts relate to real markets. The multiplier, the investment function, the consumption schedule, the wealth effect on consumption and on investment all relate to real markets. So, when there is a change in the interest rate, it takes time to affect investment. When there is a change in investment, it takes time to affect national output. When there is a change in the exchange rate it takes time to affect imports and exports. When there is a change in household wealth due to a rise or fall in stock market values or in housing prices, it takes time to affect consumption.

Real markets incorporate lags between causes and effects. These in turn can have powerful impacts on the economy.

Financial markets, in contrast, adjust with great speed. New information will cause almost instantaneous effects on the stock market, the bond market, the money market, the foreign exchange market and others. But the real estate markets tend to adjust more slowly due to institutional and regulatory impediments.

When there is a shock to the economy, the time line of its effects on the economy will be very different for real markets and financial markets. These differences then will affect the economy substantially. If there were no lags, our economy would be decidedly different.

The lags in the adjustment periods of real markets also mean that the economy is rarely in equilibrium. Usually, we are in the transition between a past disequilibrium and a future equilibrium which is always changing and will never or hardly ever be reached. So, we have to analyze transition paths, not just equilibrium positions.

# 16.

# CROWDING OUT

When the government, that great economic colossus, enters the financial marketplace to finance its deficits, it is said to "crowd out" some private sector investment by increasing the interest rate. The increase in interest rate increases the cost of financing, hence eliminating, or "crowding out", some private sector investment projects whose expected internal rate of return no longer exceeds the financing cost.

Crowding out is very incomplete. Another dollar of government debt financed expenditure will ordinarily crowd out only a small fraction of a dollar's worth of private investment. There are those who moan that "the sky is falling" (the "chicken little" effect), but in reality only a whiff of clouds usually appears on the horizon.

Indeed, the effect of government deficits is largely indirect, through its increase on total demand and, so, on the short run economy. The interest rate in turn tends to go up when the economy goes up since the demand for money rises with economic activity. This in turn results in the "crowding out" of some private investment projects which can no longer be justified due to the higher interest rate.

However, the Fed can mute, or even eliminate, the rise in the interest rate through accomodation, i.e. increase in the money supply. So we may conclude that the crowding out effect may or may not occur, may or may not be noticeable, depending on what transpires elsewhere in the economy.

# 17.

# STEEP VS. SHALLOW YIELD CURVES

The yield curve for Treasuries shows the interest rate on Treasury bills (Treasury issues with one year or less of remaining life), Treasury notes (usually considered to include Treasury issues with 1 to 10 years of remaining life), and Treasury bonds (Treasury issues of greater than 10 years life remaining).

The normal yield curve shows low interest rates for short term Treasuries (bills), higher interest rates for medium term Treasuries (notes) and still higher interest rates for long term Treasuries (bonds). The most common explanation for a rising yield curve is: risk and return. The longer the term of the Treasury issue, the greater the fluctuations over its lifetime, hence the greater the risk To compensate for the higher market risk, longer term instruments carry greater market interest yields than shorter term instruments.

However, there are other factors that affect the shape of the yield curve, notably expected future inflation over different time periods. If the rate of inflation is expected to go down in the future, then the yield curve will be flatter, since short term interest rates have to compensate more for inflation than long term interest rates. If this effect is important enough, the yield curve will be flat. If it is even more important, there will be an inverse yield curve, with short term interest rates higher than long term interest rates.

Also, if there is a "squeeze" by the Fed on short term interest rates, there is likely to be an inverse yield curve.

If the rate on inflation is expected to go up in the future, the yield curve will be steeper, with long term interest rates much higher than short term interest rstes.

In the year 2005 a mysterious development made its appearance. Just about everyone, from the Chairman of the Fed to brokerage firms, to financial advisors predicted that the long term interest rates would go up that year, as the Fed slowly increased short term interest rates a quarter point at each meeting of the Open Market Committee. But they were all wrong (as of this writing). Instead, while short term interest rates marched up steadily, long term interest rates actually declined. The yield curve flattened throughout the year.

Explanations for this unexpected development varied. But the most likely major explanation is the vast purchases of US Treasuries by China, Japan and some other Asian countries and, later in the year, by the petroleum producing countries enriched by the large increases in oil prices.The central banks of many of these countries were primarily interested in holding down the value of their own currencies in order to propagate economic industrialization and exports. They intervened heavily in the exchange rate dollar markets, purchasing vast amounts of dollars which were then converted into US Treasuries of all maturities including long term obligations. Essentially their overwhelming priority was economic development; in order to get slightly higher yields, they were willing to pay the potential price of ignoring the risk of holding long term US Treasuy obligations.

This intervention by foreign central banks was the counterpart of of the huge US import surplus which had to be financed.

Given their priorities, these central banks had to intervene. But they did not have to buy long term US obligations. They did so to get the somewhat higher yields of long terms as opposed to short terms. They seemed impervious to the risk of purchasing longer terms at a time when the short term interest rates were being pushed up steadily by the Fed.

Similar developments occurred in some other industrialized countries for much the same reasons. Financial investors, particularly some of the foreign central banks, appeared to discount risk more and more as they piled into medium and long term obligations due to their slightly higher yields. They essentially all but ignored the possibilitry of decline in the market price of longer term notes and bonds.

# 18.

# THE "TWIN DEFICITS"

It is often said that a government deficit goes hand in hand with an international trade deficit (import surplus).

However, simple observation of the facts shows that there is no such one-to-one relationship. During the '90s in the US, there were four years of Federal Government surpluses simultaneous with international trade deficits (import surpluses). In Japan, there have been many years of government deficits with simultaneous international trade surpluses (export surpluses).

There are four elements of demand in a country: (1) consumer demand; (2) business demand for investment; (3) net government demand (government expenditures minus tax receipts, i.e. the government deficit or the government surplus); (4) net export demand (an export surplus) or minus a net import demand (the import surplus, i.e. the international trade deficit).

Since there are four different parts to total demand for a country's goods and services, there cannot be a simple one-to-one relationship between any two of them. (There is no reason why the other two should remain constant.) Hence there is no set relationship between a government deficit and an international trade deficit.

# 19.

# THE EFFECT OF INTEREST RATE DIFFERENTIALS ON CAPITAL MOVEMENTS BETWEEN COUNTRIES

Often we read in the newspapers, when the Fed increases the US short term interest rate, that the increase will attract foreign financial capital flow into the US.

In a flexible exchange rate system, exchange rates fluctuate minute to minute, hour to hour, day to day, month to month and year to year. So, a foreigner who brings funds into the US at today's dollar exchange rate (the spot rate) does not know what the future exchange rate will be when he wishes to bring these funds back to his own country.

However he can buy insurance (called "buying forward cover") from a bank that guarantees the exchange rate at which he can bring back his funds to his own country at a stated future date. This rate is called the forward rate. There are different forward rates for different dates, for example the three month forward rate, the six month forward rate, the one year forward rate. The forward rates for the major currencies are published in the press.

There is a simple relationship between forward rates and spot rates for any two currencies: the difference between the spot rate and the forward rate is exactly equal and opposite to the interest differential between the two countries for the same period, with the lower interest rate country's currency trading at a forward premium. For example, with Japanese interest rates lower than US interest rates, the yen will trade at a forward premium, i.e. the forward yen will be more expensive in dollar terms than the spot yen.

This leads to the following conclusion: when financial capital moves from one country to another in a flexible exchange rate system, with free flow of capital, whatever is gained from an interest rate differential is lost due to the difference between the forward rate and the spot rate.

Hence, there is no incentive to move financial capital from a lower interest rate country to a higher interest rate country.

What happens if forward cover is not bought? Then the foreigner bringing financial capital into the US is taking his chances on the actual future exchange rate. He could lose a lot or gain a lot. However, the forward rate is the best estimate of the actual (unknown) future exchange rate. So, someone not buying forward cover will, on the average, still have to deal with the forward rate, but at greater risk. There will still be no incentive to move financial capital from a low interest country to a high interest country.

But...there is another matter that has to be considered. Sometimes, an interest rate increase will act as a signal, say, of an expanding economy and expectations of a rising stock market or real estate market. Expectations of rising asset market values do cause international capital flow in order to participate in the expected capital gains. That, in turn, can increase the spot exchange rate (and the forward rate) of the recipient country.

It should  be emphasized that it is the change in expectations that is the causal element, and changes in interest rates the signal. The cause is not the international interest rate differential itself. In fact, casual observation shows that there are interest rate differentials between countries all the time. They can persist for decades.

# 20.

# SMALL COUNTRIES AND LARGE COUNTRIES

Large countries sometimes can follow economic policies that will not work or will work poorly in small countries. And small countries can sometimes follow economic policies that will not work or will work poorly in large countries.

Fiscal policy - change in national income and output through increases or decreases in taxation and government expenditures to affect national output - will work in large countries. For example, an increase in total national demand as a result of an increase in government expenditures and/or a decease in taxation has often been used in large countries to stimulate the economy. But this will not work in small countries because much of the increase in demand will spill outside the country in the form of increased imports. So, Luxembourg cannot do what Germany can do.

A small country, such as Ireland in the late 1990's and in the early 21st century, operating in the Eurocurrency system, attracted much foreign capital and demand by revising its taxation, government expenditure and regulation system to make it more attractive to foreign business. As a result, Ireland moved from backward to boom conditions after altering its policies. A large country, like Germany, cannot do this to anywhere near the same relative degree, since capital coming from outside the country would be a much smaller fraction of its GDP

# 21.

# OVERVALUATION OF THE DOLLAR

The US dollar has been overvalued internationally for years. Largely as a result, US imports were more than half as much again as US exports by 2005. The trade deficit exceeded six per cent of our GDP by that year.

When is a country's balance of payments in equilibrium? There is a simple answer: when there is no net intervention by central banks in the international market for that currency.This does not mean that imports would be equal to exports. It does mean that the trade account in the balance of payments has to be equal and opposite to the private sector capital account. In the case of the US under current conditions that would mean that we would still have an import surplus, since private sector foreigners, on balance, want to acquire US assets.

With our currently more than six per cent of GDP import surplus, our demand for foreign currencies to pay for imports exceeds the supply of foreign currencies to pay for our exports. So, there is a very large excess demand for foreign currencies to pay for the import surplus. Alternately, we can say that the supply of dollars to pay for our imports exceeds the demand for dollars to pay for our exports. So, there was a large excess supply of dollars. Since this imbalance was enormous, it would by itself have brought drastic decline in the value of the dollar, i.e. increase in the value of foreign currencies. In fact, the average value of the dollar did decline substantially, but not drastically, from 2003 to the early part of 2005. (For the rest of 2005 the dollar rose in value due to short run reasons.)

What sustained the value of the dollar? The answer lies in net capital imports, the net sale of US assets to foreigners. Since acquisition of US assets requires dollars, the net acquisition of US assets generated net demand of dollars on capital account.

The demand for US assets arose from two sources, the private sector and foreign central banks. Demand on private account was generally not enough to plug the hole left by the trade deficit. So, foreign central banks essentially made up the difference. The US balance of payments was not in equilibrium.

Some foreign central banks were required to intervene in the dollar market; most were not required but chose to intervene anyway in order to prevent their own currencies from rising too much relative to the dollar.

The Chinese central bank in particular was required to intervene to maintain its fixed official exchange rate and, later, its almost fixed exchange rate. China acquired huge amounts of US assets, mainly US Treasuries. Many of the other central banks, particularly Asian central banks, notably Japan, chose to intervene (although not required to do so), also acquiring large amounts of US assets, again mainly in the form of US Treasuries. When oil prices rose steadily, the oil producing countries also became prominent in acquiring US assets.

So, it was foreign central banks' acquisition of dollar denominated assets that held up the value of the dollar. Without such interventions, the dollar exchange rate would have gone down greatly.

# 22.

# TYPES OF EXCHANGE RATES

There are many different kinds of exchange rates: spot rates and forward rates; nominal bilateral rates and real bilateral rates; nominal multilateral rates and real multilateral rates.

When we look in the newspaper, the exchange rates given for any US/foreign country combination are nominal bilateral spot rates and nominal bilateral forward rates. The spot rates are for delivery of a currency "on the spot". The forward rates are for delivery of a currency at some time in the future, for example in 3 months or 6 months. Both spot and forward rates can be given in either directions, e.g. euros per dollar or dollars per euro.

There is a relationship between spot rates and forward rates. The difference between these two is exactly equal and in the opposite direction to the difference between the interest rates in the two countries for the same time period. The country with the lower interest rate will trade at a forward premium, i.e. the currency of that country will be more expensive when desired in the future than in the present. The currency of the other country will trade at a forward discount.

Bilateral nominal exchange rates are not adequate when two countries are compared over a period of years. For that, we need bilateral exchange rates that have been adjusted for the differing rates of inflation in the two countries, i.e. bilateral real rates.

When a country's position is compared to that of its trading partners as a whole, we need multilateral exchange rates. Again, this can be done in nominal terms or in real terms. Basically, the bilateral exchange rates of a country are weighted by the importance of each trading partner in terms of exports and imports and the multilateral exchange rate is calculated. Often only the top 5, 10, or 15 trading partners are included to simplify the mathematics and the data acquisition. In making comparisons of a country's exchange rate over a period of years we need the multilateral real exchange rate.

All this is true in a flexible exchange rate system. This has been the system for the major countries since 1973 and, to some extent, since 1971. The old fixed exchange rate system that was evolved in the Bretton Woods agreement of 1944 for the postwar period collapsed in 1971, was partially resuscitated subsequently, and completely collapsed in early 1973.

# 23.

# FIXED VS. FLOATING EXCHANGE RATES

There are two major exchange rate regimes: fixed and floating.

Today, the exchange rates of the major developed countries (or country systems) - the dollar, the euro, the yen, the British pound - are floating relative to each other. The major exception is China which had maintained a fixed rate, and later had an almost fixed rate, relative to the dollar. And the euro system, encompassing most countries of Western Europe, has only a single currency now, the euro.

In a floating exchange rate system, the foreign central banks may or may not intervene in the international dollar market. Since the US has had a whopping big import surplus, and since private sector US net capital imports have often not been sufficient to supply all of the foreign currencies needed to pay for the US import surplus, foreign central banks often intervened to supply the remaining US needs for foreign currencies. The question becomes this: at what level of the dollar exchange rate do they intervene? If they intervene to sell their own currencies and buy dollars with the purpose of keeping their own currencies down in value, then the value of the dollar will be high. If they are willing to let their currencies rise, then the intervention point will be at a lower value of the dollar.

In effect, the intervention policies of foreign central banks will largely set the exchange value of the dollar in a flexible exchange rate system, if net US private sector capital imports are not sufficient to cover the US import surplus. (Note: the US central bank, the Fed, rarely intervenes.)

But, in a broad sense, there is one significant limitation to intervention by foreign central banks. A foreign central bank can always sell its own currencies to buy dollars in order to keep down the value of its own currency. But it cannot always do the opposite, sell dollars and buy its own currency , in order to keep up the value of its own currency. To do this, it must have (or borrow) dollar reserves. If it runs out of reserves, its currency will drop, witness the Asia crisis of 1997-98, starting with the collapse of the Thai baht.

If a country wishes to maintain a fixed exchange rate relative to the dollar, it must intervene all the time; otherwise, the fixed exchange rate will become unfixed. So, the Chinese central bank must intervene constantly in the dollar market to keep the Chinese currency, the renmimbi or yuan, fixed or almost fixed relative to the dollar. Since the yuan is undervalued, the Chinese central bank must stand ready to buy dollars. Otherwise, the yuan would rise substantially relative to the dollar.

What can we conclude under current world conditions: first, the dollar is overvalued; it requires a huge amount of borrowing from other countries to maintain its value. Second, the Chinese yuan is undervalued. It requires a large amount of lending to maintain its low value, and the dollar's high value.

# 24.

# DIFFERENT GROWTH RATES IN DIFFERENT COUNTRIES

Long term growth rates vary tremendously in different countries. Some have had almost double digit real growth rates at times. Others actually have had negative growth rates for some periods. What explains these vast differences?

In order to grow, a country must set aside a portion of its GDP to increase the quantity of the factors of production. Increase in the factors of production, in turn, increases future output.

A complicating, and negative, factor for economic growth is the rate of increase in the population and in the labor force. Part of what is set aside to increase the factors of production must be used to take care of population and labor force growth.

The portion of the GDP set aside as gross investment is used for: (a) replacement investment due to depreciation; (b) "widening" of capital to equip the increase in the workforce with the same capital per worker that the preexisting workforce already has; and (c) "deepening" of capital, i.e. increase in the capital per worker of the existing workforce. The greater is the increase in population and workforce, the more of the setaside will be used for that purpose, and the less will remain for increasing the ratio of capital to labor needed for economic growth.

Similarly, a portion of GDP has to be set aside to produce increase in human capital, i.e. economic skills through education. Here too, the greater the increase in population and and in the labor force, the less will remain to "deepen" the education and skills of the existing population and labor force.

There is, as well, the highly important matter of technology. Either technology will have to be acquired from elsewhere, or it will have to be self produced. In either case, part of national output will have to be set aside to acquire technology.

For the advanced, highly developed countries, this means acquisition of newly created technology internally or from other highly developed, advanced countries. These countries have already exploited old technology. Hence they need new technology in order to grow. Research and Development creating new technology is a world phenomenon principally originating in the thirty or so advanced countries. Without such new technology, economic growth would eventually stop. New technology is the engine of economic growth in the advanced countries

For the rest of the world, the underdeveloped countries and countries at some middle stage of economic development, technology that is "old" to the advanced countries but "new" to them will be used in large measure.

It should be noted that most, but by no means all, technology is incorporated in capital goods, both in net investment and in replacement investment.

Poor countries have great difficulty in setting aside any part of their GDP to increase the factors of production per worker. And poor countries tend to have high population growth rates, making it even more difficult to achieve growth. To some extent, foreign investment used for capital formation can help. But, many poor countries have low or even negative growth rates per capita.

For those countries that succeed in setting aside enough of their GDP to get growth going, conditions during the growth process itself are likely to change with the passage of time. Population growth rates eventually go down, investment rates go up, educational investment rates go up, absorption of technology rises. As a result, the growth rates often become high for a fairly long period of time. The Asian tigers - Japan, South Korea, Singapore, Taiwan, Hong Kong - are good examples, as is Israel. It seems that a new set of tigers - probably including Thailand, Malaysia, Viet Nam - are likely to follow. But eventually, after some decades, growth tends to slow as these countries come closer and closer in terms of GDP per capita to the highly developed, industrialized countries like the US and most of Western Europe

So, growth rates vary both geographically and in time,depending on the stage reached by countries in the developmental process.

# 25.

# CONVERGENCE

One of the profound questions of our age is whether labor productivity and living standards in different countries will converge in a few decades to those of the leading economies such as the US and Western Europe.

Convergence or lack of convergence is very difficult to predict since the time periods concerned are very long - decades, half centuries and even centuries. Who would have predicted in 1950 that South Korea and Taiwan (both former colonies of Japan), Hong Kong and Singapore (British) would be highly successful in economic growth and essentially near converge to the living standards of the leading countries today. And, of course, there was Japan.

Today, it is equally difficult to predict. But Thailand, Malaysia and Viet Nam look like they are on the way to success. So does China.

There are vast diffferences in per capita incomes around the world, on the order of 25 to 1 or even more. Both the relative and absolute differences are enormous.

Looking back 2 1/2 centuries, in the year 1750 both the absolute and relative differences were a lot less. At that time, just about everybody was poor, even though the degree of poverty varied substantially. But, since that era, some countries have advanced mightily and others have lagged behind or even stagnated completely. So, today the relative and absolute differences are far greater.

If history is a guide, only a modest number of countries will converge to the living standards of the industrialized countries in the next half century. Others will advance so that their incomes will rise relative to the leading countries, though absolute differences may actually grow. Still other countries, mostly currently poor countries, will advance very little, so that both relative and absolute differences are likely to rise.

What factors will be important in rapid economic growth? Vital is the proportion of the national income set aside to increase the factors of production. Gross investment has to rise to increase the capital stock. More of the national product has to be devoted to increase the stock of education. More has to be utilized to pay for imported technology.

At the same time, population growth will decline with rise in living standards. The high rate of population growth in many underdeveloped countries, particularly in Africa, is a serious impediment to development, since the increase in population and labor force eats up part of the setaside to expand the factors of production relative to labor.

It turns out that other factors are also important, particularly climate and access to the sea. Hot climates come with tropical diseases and often high, enervating humidity. Lack of ocean access multiplies transportation costs to foreign markets and from foreign sources.

So, will all countries converge? No. Will at least all countries gain in relative terms on the highly successful industrialized countries? No. Some will, others not. We can conclude that, a hundred years from now, more countries will join in high living standards and converge. Others will climb part of the way. But for a great many, relative and absolute differences with remain much the same or even worsen.

# 26.

# ECONOMIC GROWTH, PER CAPITA GROWTH AND GROWTH IN LEISURE

We often read in the newspapers that country A has a growth rate of 4% and country B only 3%; so, country A must be doing better.

Such a conclusion is fallacious.

It fails to take into account different population growth rates.

If country A has a population growth rate of 1% (call it the US) and country B has a population growth rate of zero (call it much of Western Europe), then the per capital growth rates are actually equal.

There is also the matter of leisure. Some countries choose more leisure time and less work, others more work and less leisure time. US living standards are higher than Western European living standards because Western Europens work fewer hours per year. As a whole, they retire earlier and have much longer vacation periods plus holidays. This difference in living standards holds as long as we only measure actual output per capita.

But, when we decide to value leisure, the difference in living standards between the US and Western Europe vanishes; the US and most of Western Europe then have roughly the same living standards.

If we look at this historically, it is obvious that all advanced countries have taken the result of advances in technology over the past two hundred and fifty years to a substantial degree in more leisure time and less work. We used to work six days a week, up to 12 hours a day or even more, with very little in the way of holidays and vacations. Now we work 5 days a week and perhaps 8 hours a day or less with far more time off from work.

We can also take into account both leisure and population growth rates directly into growth calculations.

If a country A has a 4% growth rate in measured output, with a 1% population growth rate with no change in per capita leisure, while a country B has a 2% growth rate in measured output, with a 0% growth rate in population, but a 1% per capita reduction in annual hours worked, then the per capita growth rates of the two countries are actually the same, provided that we put a value on leisure.

So, it is important in per capita growth rate comparisons between different countries to: (a) allow for differences in population growth rates and (b) allow for differences in the reduction rate (negative growth rate) of annual hours worked.

# 27.

# LABOR PRODUCTIVITY GROWTH

Labor productivity growth, the growth rate of output per worker or per worker hour (but the two are not identical), is extremely important , since the growth rate of long term living standards is determined by the growth rate of long term labor productivity.

The growth rate of GDP can be divided into two components: the growth rate of labor productivity and the growth rate of the labor force. These two must be added to get the growth rate of GDP.

For the highly developed, industrialized countries, the growth rate of labor productivity essentially depends of the growth rate of of technology, i.e. technological progress.

Technological progress in these countries in turn depends on newly developed technology, since the older technology has already been utilized in the past. In the last decade or so, particularly after 1995, technological change has taken an upward jump, a favorable development for long term expansion in living standards. We do not know whether this greater rate of technological progress will continue indefinitely. At the present time the outlook is favorable.

For the successful countries in the developing world, the growth rate of labor productivity can be much higher, and so can the growth rate of GDP. These countries can take advantage of the available existing technology (which is new to them, but old to the highly developed, industrialized countries), as well as the newly created technology.

They can also benefit from the rise in the capital/labor ratio, the education/labor ratio, the shift out of the less productive agricultural sector into the growing more productive industrial sector, and shift out of the poorly productive services into manufacturing. As a result, their living standards can expand rapidly.

It should be noted that, for the highly developed industrialized countries, the only real way to increase hourly labor productivity is through technological change. New technology, by definition, is not subject to the law of diminishing returns.

But, in principle, the other factors of production are subject to this famous law. Hence, increasing the capital/labor ratio or the education/labor ratio by itself would run into the law of diminishing returns eventually. Of course, in practice, most new technology is incorporated in capital equipment, including both new and replacement capital equipment.

# 28.

# CAPITAL GAINS, INFLATION AND TAXES

A capital gain is the difference between the buying price and the selling price of an asset, such as a stocks, bonds or real estate.

But, there may or may not be a real capital gain (as opposed to a nominal capital gain), depending on the degree of inflation during the holding period of the asset.

We may use an example. If an asset is held for a year and a day, with a capital gain of 3%, at a time when inflation during the period was 4%, then there is a nominal capital gain of 3%, but a real capital loss of 1%.

However, the US taxation system is not adjusted for inflation. So, with a long term capital gains tax of 15%, the after-tax nominal capital gain is only 85% times 3%, or 2.55%. The real capital loss is 1.45%.

If the asset is held for more than a year with a capital gain of 3%, with an inflation rate of 3%, the nominal gain is zero. However the after tax picture is even bleaker. With a 15%

capital gains tax rate, the real capital loss is 0.45%. In effect, the government is taxing the rise in the price of the asset due simply to inflation.

This feature of the US tax system, taxing non-existing real capital gains, has long elicited complaints, but to no avail.(Fewer complaints are heard today, since the tax rate on capital gains is much lower than it used to be.)

However, taxes have to be paid only when the asset is sold. So, assets can appreciate for many years without triggering a tax - at least if the asset is directly held, instead of indirectly (as in a mutual fund). That is a distinct advantage. And, if the asset is held until the death of the owner, there is no capital gains tax at all. Instead, an asset is written up to its new value when it becomes part of the estate of the deceased owner, and is only subject to whatever estate tax is due, if any.

# 29.

# BUBBLES

When the valuation of assets, such as stock market assets or private houses, reaches "unreasonable" levels. we call it a bubble

Since one man's bubble is another man's highly valued asset, there is no universal agreement on the designation of particular situations as bubbles.

Still, there is at least substantial agreement in certain instances. The Dutch tulip mania in the 17th century is widely thought to be a bubble. The Japanese real estate market in the 1980's is also thought to qualify as a bubble. So is the American stock market at the end of the 1990's, particularly the NASDAQ market and the dot com companies. Also included are some local real estate housing markets as of 2005.

Bubbles can be perfectly rational. If I believe that a certain asset is not worth its current market price, but also believe that it will rise in price, I will buy that asset, even though I think it is part of a bubble. I buy because I expect to sell at a higher price to a greater fool.

A great deal of money can be made in a bubble, provided that it is possible to get out before the collapse. A great deal of money can also be lost in a bubble by those who buy high and fail to sell before the collapse.

Isaac Newton was deeply involved in the South Sea bubble. At first, he bought low and sold higher. He did well. But then, as prices continued to rise, he got back in and was caught in the inevitable eventual collapse. Timing is everything in bubbles.

72

# 30.

# INCOME DISTRIBUTION

Income distribution is an important subject. But its near absence in economics texts is notable, whether caused by lack of data, lack of interest, or lack of analytical explanation.

There are at least two important types of income distribution. The first is functional income distribution, the share of the national income going to wages and salaries on the one hand and the share of national income going to owners of capital assets on the other.

A commonly used figure is around two-thirds as labor income and one third as capital income. In the US, the labor income proportion has been closer to three-quarters, particularly when the labor component of small business is factored in. That leaves one-quarter to owners of capital assets.

It should be realized that labor income and capital income overlap for many individuals and households, since people both work and receive income derived from ownership of capital assets.

That brings us to the second type of income distribution, namely personal income distribution. Data typically come from income

Peter Gutmann

tax returns. This means that the definition of income includes capital gains. So, this type of income distribution is somewhat different from the income distribution that might, at least in principle, be derived from national income account data alone (since capital gains are not included in the definition of income in the national income accounting system).

Income distributions tend to be skew, with a relatively small number of very high income recipients at the top and a large number of relatively low income recipients at the bottom.

Such income distributions are important in determining the national savings ratio. The bottom half of the income distribution usually does very little saving. The top half, and in particular the top few percent, does practically all the savings. Indeed, the lower income groups actually do a lot of dissavings.

In the US a great deal of information on income distribution has become available in recent years. Briefly, as of 2001, the top one-tenth of one percent of income recipients (one in a thousand) ( receiving income of $1,589,608 or more, 2005 dollars) received 8.0% of the national income; the top one-half of one percent (one in 200) ( receiving income of $581,019 or more) received 13.9 per cent of the national income.

The top one per cent (one in a hundred) (receiving income of $383,407 or more)  received 17.7 per cent; the top five per cent of income recipients (one in twenty)   (receiving $162,351 or more) received 32.7 per cent of the national income.

The top ten per cent of income recipients (one in ten) (receiving $117,001 or more) received 43.9 per cent of the national income; the top 20 per cent, i.e. the top quintile, or one fifth (one in 5) (receiving income of $79,562 or more) received  59.7 per cent of the national income.

The bottom one-fifth of income recipients (one in 5) (receiving income of $13,478 or less) received just 2.5% of the national income. The bottom two-fifths (two in 5)  (receiving income of $25,847 or less) received 8.9% of the national income. The bottom three-fifths (three in five) (receiving income of $44,451 or less) received 20.6% of the national income, The bottom four-fifths (four in five) (receiving income of $79,562 or less) received 40.7% of the national income). (Figures subject to rounding)

So, the top one-fifth of income recipients got nearly 60 per cent of the national income, while the bottom four-fifths of income recipients got about 40 per cent of the national income.

The top one-tenth of income recipients got somewhat more of the national income than the bottom four fifths of income recipients. These are the facts.

Taxation of income can and does affect after-tax income distribution. Taxation of higher incomes at higher rates and lower income at lower rates will make the after-tax income distribution less skew. This is called a progressive income tax system. The difference between pre and post tax income distribution depends on the progressivity of the tax system.

In the US, the Federal tax system (including both income tax and social security tax as well as other Federal taxes) is progressive up to around the 90th percentile, but then basically ceases to be progressive. In other words, the total tax rate including all Federal taxes (i.e. the proportion of income that goes to the Federal Government as taxes) at the 90th percentile is very little different from that tax rate at the 99th percentile.

In addition, there are state and local government taxes, particularly the sales tax. These are also significant in determining after-tax income distribution. It is well known that sales taxes are regressive, i.e. they hit the lower and middle income groups hardest, since the proportion of their income spent on consumption goods and services is higher among low and middle income groups than in high income groups. So, the presence of sales taxes will reduce the after-tax progressivity in a progressive income tax system. Much the same conclusion can be reached for the European value added tax system.

Peter Gutmann

The vast complexity of taking into account not only Federal taxes of all sorts but also sales taxes, real estate taxes, etc. means that after-tax income distribution including the whole panoply of Federal, state and local taxes would be exceedingly difficult to calculate and thus has not been available.

# 31.

# WHO PAYS FEDERAL TAXES IN THE US

There is a considerable amount of confusion about the allocation of the Federal Government tax burden to different income groups in the US. This appears to be mainly due to the publication of charts and tables showing the allocation to different income groups of the personal income tax alone (excluding the Social Security tax and other Federal taxes).

The personal income tax is only somewhat more than two-fifths of total Federal taxes. The Social Security payroll tax (including the Medicare tax) the next largest in terms of receipts, is just a bit smaller than the personal income tax. Including the payroll tax share paid by employers ( which economists agree comes out of wages and salaries), four in five employees pay more in Social Security (including Medicare) tax than in income tax.

Including all Federal taxes - income tax, Social Security (including Medicare) payroll tax, excise taxes on liquor, tobacco, etc.(2004) - the top 1 per cent of income earners pay 23% of Federal taxes, the top 5 percent of income earners pay 40% of Federal taxes, the top 10 percent pay 52% of Federal taxes, the top 20 percent pay 69%, while the bottom 80 percent pay 31% of Federal taxes.

In terms of the percent of income taken by all Federal taxes (2004), the top 1 per cent of income recipients pay 26.7% of their income, the top 5 percent of income recipients pay 25.6 percent, the top 10 per cent pay 24.9 percent and the top 20 percent pay 23.8%.

The next quintile (one-fifth) of income recipients pay 18.5% of their income, the third quintile of income recipients pay 14.6% of their income, the fourth quintile of income recipients pay 11.1% of their income and the lowest quintile of income recipients pay 5.2% of their income.

It should be noted that the progressivity of the Federal tax system, going up the ladder from the lowest quintile to the fourth quintile, keeps on rising. But there is very little   progressivity within the top quintile.

Once state and local taxes are included, particularly sales taxes, the progressivity of the  total tax ystem is reduced, since sales taxes (also value added taxes) are regressive.

Another point important in the tax system is the source of income. The percentage of income from capital gains and dividends rises from 3% for adjusted incomes ( year 2000) for those under $50,000 to 61.4% for those over $10 million. Since long term capital gains (held for more than a year) are taxed at a lower rate, this has significant bearing on the total Federal tax burden of different income groups.

# 32.

# SOME FACTS ON US INFLATION IN THE "COST OF LIVING"

Inflation rates compound over time. For example, 10 per cent inflation means 21% inflation over two years. This compounding makes it somewhat difficult to calculate the total degree of inflation over an extensive period of time.

Here are the results of total inflation over a decade, based on stated annual inflation rates: 2 per cent annual inflation compounds to 22% in 10 years; 3 per cent annual inflation compounds to 34% in 10 years; 4 per cent annual inflation compounds to 48% in 10 years; 5 per cent annual compounds to 63% in 10 years; 6 per cent annual inflation compounds to 79% in 10 years; 7 per cent annual inflation compounds to 97% in 10 years. Finally, 10 per cent annual compounds to 159% in a decade.

What about the actual inflation in the "cost of living" (the consumer price index) over the decades in the post World War II era in the US? Here are the results.

From 1945 to 1955 total inflation was 49%. From 1955 to 1965 it was 18%. From 1965 to 1975 it rose to 71%. From 1975 to 1985, with double digit inflation rates for some years, the decade total jumped to 100%. After that less than satisfactory experience, the next decade, 1985 to 1995, saw a diminution of inflation to 42 per cent for the decade, particularly due to Fed action in restraining the growth rate of the money supply in the early 1980's. Finally, for the most recent decade, from 1995 to 2005, the total was only 31%.

For the period as a whole, the sixty years from 1945 to 2005, inflation totaled 1007% per cent. Few people realize that this number is so big. Prices are now at not far from eleven times the 1945 total.

For the even longer period of 1913 to 2005, approaching a century, inflation totaled 1,912%, i.e. prices multiplied by about twenty times.

It is obvious that prices rose at a considerably faster rate in the period after World War II than in the roughly three decades before World War II. One reason, but not the only one, is the negative rate of inflation (i.e. deflation) in several early Great Depression years.

This difference between pre World War II and post World War II is a well known phenomenon. One result is that far more attention has had to be paid to inflation after World War II than in prewar years . And far more effort has gone into the prevention of inflation.

# 33.

# SOME FACTS ON US PER CAPITA GROWTH RATES IN REAL GDP

Growth in real GDP per capita is extremely important because it is necessary to secure long range increases in living standards.

Growth rates compound over the years. To put the US figures in perspective, a 2 per cent annual growth rate compounds to 22 per cent in a decade.

Here are the results of US growth rates in real per capita GDP per decade over various decade time periods.

From 1950 to 1960 it was 18.2%. From 1955 to 1965 it was 22.7%. From 1965 to 1975 it was 21.4%. From 1975 to 1985 it was 26.8%, From 1985 to 1995 it was 18.2%. From 1992 to 2002 it was 22.5%. These figures are affected somewhat by the short run state of the economy in the base particular years chosen. But clearly, the growth rate in real GDP per capita averaged around 2 per cent per year during this period.

For the 52 years from 1950 to 2002, the total growth in real GDP per capita was 195 per cent. In other words, real GDP per capita almost tripled over slightly more than half a century.

For the even longer period of 1929 to 2002, a period of 73 years, the real GDP per capita advanced 387%, i.e. it almost multiplied by five.

For a highly industrialized country, dependent on further advance of new technology, the US has had a good past half century.

# 34.

# SOME FACTS ON US POPULATION GROWTH RATES

Growth in population is important. Additional population creates additional demand as well as additional supply. It is particularly important in the housing market, since net additional demand for housing is added, due to population increase, to the existing replacement demand and upgrading demand.

To put the US growth rates by decades given below in perspective, here are some annual growth rates and the result of compounding over 10 years. A 0.9% annual growth rate compounds to 9.4% total for a decade. A 1.0% annual growth rate compounds to 10.5% for a decade. A 1.1% annual growth rate compounds to 11.6% for a decade. A 1.2% annual growth rate compounds to 12.7% for a decade. A 1.3% annual growth rate compounds to 13.8% for a decade. And, finally, a 1.7% annual growth rate compounds to 18.4% for a decade.

The actual figures for the US follow. From 1950 to 1960, population rose 18.5%. From 1960 to 1970 population rose 13.4%. From 1970 to 1980, population rose 11.4%. From 1980

Peter Gutmann

to 1990 population rose 9.8%. From 1990 to 2000 population rose 13.1%.

In recent years, roughly one-third  of the population growth has been due to net immigration. The other two-thirds of the population growth was the result of the difference between birth rates and death rates in the US. (The US death rate is roughly three-fifths of the US birth rate.)

# 35.

# BEHAVIOR AND THEORY

A theory is simply an abstraction of reality. Reality is complex. Theory allows us to cut through the complexities sufficiently so that we can see the forest, not just the individual trees.

And, since much of economics deals with behavior, theory must deal with the realities of behavior. It must deal with what people do, not with what people supposedly ought to do based on some economist's assumptions about behavior.

There are all kinds of theories in macroeconomics that do not work in the real world. For example, there is a well known theory that states that private sector savings will go up as a direct result of a rise in the government deficit, and will go down as a direct result of a fall in the government deficit. This theory assumes that people will behave in such a manner that they will save more if the government deficit goes up because they then assume that taxes will go up later to pay back the resulting government debt. Unfortunately this theory doesn't work in practice because people do not behave in such a manner in reaction to a rise in the government deficit.

There is another theory - much in vogue these days, especially among a substantial number of politicians - that private sector savings will go up directly if the deficit goes down. Actually, a change in the government deficit, up or down, will result in a complex series of interactions that affect the whole economy, and with it, the magnitude of private sector savings. The most likely short run results of a decline in the deficit alone will be a decline in the economy and a decline in private sector savings.

Another theory holds that a rise in the government deficit causes a rise in interest rates, and a fall in the government deficit causes a fall in interest rates. But, the relationship is not between interest rates and the government deficit, but between interest rates and total economic activity of the country through complex interactions. So, when government deficits go up, interest rates may go up or down, depending on general economic conditions. When government deficits go down, once again interest rates may go up or down.

Then there is the theory that a decline in income tax rates will actually increase government tax revenue via a large incentive effect through savings and investment that increase the supply of output sufficiently. This is an extreme statement of results that hold only under very stringent behavioral assumptions and very high income tax rates. Incentives do matter of course. Indeed they are very important, but a theory must not assume exceedingly powerful effects that are unlikely to occur under current US conditions.

So, the major point is this: theory must correspond to reality. If it does not, as evidenced by actual results of the theory's application, since the facts are not going to change to accord with the theory, the theory must be changed to accord with the facts. Theories are always being modified. Some have to be abandoned.

# 36.

# POLICY: THE SHORT RUN AND THE LONG RUN

Government policies appropriate to the short run are often inappropriate for the long run.

The short run is a period short enough to reasonably assume that the quantities of the factors of production and the output capacity of the economy are unchanged. In the long run, the quantities of the factors of production and the output capacity of the economy change.

Monetary policy, by its very nature, tends to be short run policy. The Fed controls short run nominal interest rates. These can vary substantially from one year to the next. A major Fed objective is control over short and long run inflation. The Fed influences, but does not control, long term nominal interest rates which are affected by expected inflation, risk and return characteristics and other factors. The Fed tries to influence long run interest rates by signaling its future intentions through statements issued after meetings of the Open Market Committee, later release of the minutes of these meetings, and speeches and Congressional testimony by its chairman and members.

The major problems arise in fiscal policy. Here, good policies in the short run are often poor policies in the long run and vice-versa.

In a recession, for example, a good policy may require tax reductions and government expenditure increases in order to stimulate the economy through increase in total demand for domestic goods and services. And...tax decreases have to be presented as "permanent" since it is well known that temporary tax reductions have a much lesser impact in boosting the economy than permanent tax reductions. But all this is likely to create a substantial government deficit.

A few years down the line, when the economy has recovered, and when other conditions may well have changed - for example the size of medical program costs and pension expenditures - continuation of the short run policy adopted at a time of recession becomes inappropriate. And so, there then have to be "permanent" tax increases and/or expenditure decreases. This, in fact, happened in the US in the 1980's.

The same principle applies in reverse to boom times when inflation threatens. This is a time to increase taxes "permanently" and reduce government expenditure in order to reduce total demand for domestic goods and services. But...this short term policy becomes inappropriate a few years later when the economy is once again in recession and another set of "permanent" changes is required in taxes and government expenditures.

An obvious difficulty lies in the use of the concept "permanent". The more people realize that "permanent" lasts only a few years, the less will such changes affect the economy and the more difficult it will be to use fiscal policy.

The contrast netween the short run and the long run has been known for many years. No business will adopt short run policies and keep these unchanged forever. But government is elephantine, lumbering and hard to move. So, changes come erratically at unspecified intervals. In the meantime, inappropriate policies - which once were appropriate - can continue much too long a time.

# 37.

# THE ACHILLES HEEL OF THE US ECONOMY

There are many serious problem areas in the US economy. Medicaid, medicare, Social Security, pension benefits, expenditures for prisons etc. - all are substantial problems that will have to be addressed sooner or later. Some are short and medium term problems, others long term problems. In all of these, government expenditures have been rising. In some, expenditures cannot be sustained at recent rates of expansion in the long run. In some, the Federal Government faces the problem (e.g.Medicare, Social Security and pension benefits), in others the States are more involved in the problem (e.g. Medicaid and prisons).

But the most significant problem of all, short and medium run, is the staggering import surplus, the enormous trade deficit, that is running at more than 6 per cent of US GDP currently and rising. This is a macro problem par excellance.

It has several aspects. First and foremost, the huge import surplus has the potential of sending the economy into recession, quite possibly serious recession.

A prosperous economy requires that enough total demand exist to purchase the supply that the economy produces when operating at high levels. There are four parts to total demand: consumer demand, business investment demand, government demand and net foreign demand. But in the US one of these, net foreign demand, is an enormous negative, namely the more than 6 per cent of GDP import surplus, the excess of our demand for imports over the demand for our exports. This excess is demand that spills out of the country. It subtracts from the demand for domestic goods and services.

This net negative demand must be compensated by the other three parts of total demand.

Consumer demand: this, the largest of the four elements of total demand, comprising about three quarters of the total, is already stretched to the utmost. Household savings have lingered near the zero level for a number of years, indeed have been negative in some periods. The upward push to consumer demand from the remortgaging boom, that permitted homeowners to take out more and more cash as house prices rose, has hit top and will be declining. A good deal of the remortgaging was financed by short term interest rates which have been rising. Mortgage rates are also up somewhat. So, it is not very likely that household consumer demand will rise much further (as a per cent of the GDP).

Government demand: the Federal Government has already been running a substantial deficit. There is an important sector of opinion that wants to reduce this deficit. Efforts have been under way in Congress to accomplish this objective through reduction in government programs. (These may or may not succeed.)

The Federal Government deficit pumps net demand into the domestic economy. To some extent, this has compensated for the shortfall in demand due to the import surplus. But realistically speaking, considering the size of the existing government deficit, it does not seem at all likely that there will be a push to increase the Federal deficit (as a proportion of of the GDP) as a way of compensating for any increase in the import surplus (as a percentage of the GDP).

Business investment demand: that leaves the last of the four parts of total demand, namely business investment demand. This element of demand has been doing well. It, too, has partially compensated for the deficit in demand due to the import surplus. Business investment demand is a net demand factor insofar as business investment exceeds business savings. But the question is this: will business investment demand increase sufficiently (as a per cent of GDP) to compensate for any further increase in the import surplus (as a per cent of GDP)?

So, at the time of this writing, total domestic demand for goods and services has been adequate to purchase total domestic supply. But, the import surplus has been rising, both absolutely and as a per cent of GDP.

What happens if the import surplus gets even bigger as a per cent of GDP? Will a rise in business investment be large enough to compensate for a further increase in the shortfall of demand as a result of the import surplus?

Most importantly, what happens if investment demand declines. This can be critical.

And, what happens if consumer demand becomes a smaller per cent of the GDP (i.e. household savings rise as a per cent of the GDP)? And what happens if the Federal Government deficit falls as a per cent of the GDP?

The answer is: recession.

GDP will have to fall (or perhaps grow more slowly) until the equality of demand and supply is restored at a lesser income level.

There is a second serious problem implicit in the huge import surplus. It involves the exchange rate of the dollar. The present level of the dollar has been sustained by large purchases of dollars by private foreign investors and, very importantly, by foreign central banks. The central banks of a number of foreign countries want to keep down the value of their own currencies so that their countries can export as much as possible and run an export surplus. This particularly involves a number of Asian countries, notably China, as well as some Middle Eastern countries. Many of these countries are in the process of industrialization and economic development and are quite anxious to keep up their manufacturing growth through export demand. So, they sell their own currencies to keep down their value and buy dollars. This keeps up the value of the dollar.

The result of these dollar purchases is enormous acquisition of US assets, particularly US Treasuries. Around half of all publicly available US Treasuries are now held by foreigners.

But...how long will private sector foreigners and foreign central banks be willing to buy an asset (dollars) whose value is being kept up only by their own purchases, an asset that is bound to decline measured in terms of their own currencies once these purchases of dollars diminish sufficiently to send the dollar's value way down?

The danger is a large decline in the international value of the dollar. Of course that decline itself would tend to decrease US imports and increase US exports. But even a large decline would hardly eliminate the US trade deficit. Nor would it achieve "balance" (a situation where net central bank intervention in the international dollar market is zero) in US economic relations with foreign countries. Both US imports and exports tend to be inelastic (i.e. do not respond very much to price). For example, in an age where even the purchase of "American made" goods, on closer look, involves many foreign made components, there is great inflexibility of import demand.

And a large decline in the value of the dollar would make imports more expensive. With our high quantity of imports, that then becomes a factor making for  more domestic inflation.

To summarize: the unsustainable enormous US import surplus has both domestic and foreign implications. Domestically, there is a huge negative demand for US output as a result of the import surplus. This has to be compensated by the other three elements of demand: net government demand, business investment demand and consumer demand. Externally, the value of the dollar is on an unsustainable path and will have to decline drastically to reduce the import surplus to sustainable levels.

# APPENDIX

Sources

Chapter 30: Data Sources described in <u>New York Times</u>, June 5, 2005, page 27

Chapter 31:Data from Joint Committee on Taxation, see <u>New York Times</u>, April 11, 2004, page 2 of section 4 <u>and</u> "Effective Federal Tax Rates Under Current Law", Table 2, Congressional Budget Office, August 2004 <u>and</u> <u>New York Times,</u> October 6, 2004 (from Tax Policy Center)

Chapter 32: US Department of Labor, Bureau of Labor Statistics, Inflation Calculator, Consumer Price Index - All Urban Consumers (see: www.bls.gov)

Chapter 33: <u>Statistical Abstract of the United States, 2003</u>, US Census Bureau, Mini Historical Statistics, Selected Per Capita Income and Product Items in Current and Real (1996) Dollars, 1929 to 2002, table number HS-33

Chapter 34: <u>Statistical Abstract of the United States, 2001</u>, US Census Bureau, Population, Table Number 1, Population and Area, 1790 to 2000 <u>and</u> Table Number 4, Components of Population Change, 1980 to 1999

Chapter 36: This chapter is a slightly modified version of chapter 29 in Peter Gutmann, <u>Understanding Modern Macroeconomics,</u> Authorhouse Publishing Company, 2005_

# About the Author

Peter M. Gutmann is professor of Economics at Baruch College of the City University of New York.

He has a doctorate from Harvard University. His dissertation title was "Income Distribution, Asset Values and Economic Growth".

Professor Gutmann is widely known due to his pathbreaking work on the subterranean, or underground, economy which created a whole industry of articles and books by economists from all over the world on that subject.

He is the author of "Macroeconomics in Brief" and "Understanding Modern Macroeconomics". He has published in a range of economic journals including the American Economic Review, the Journal of Income Distribution, the Review of Economics and Statistics, and others.

Professor Gutmann teaches macroeconomics and growth economics at Baruch College of the City University of New York.

www.ingramcontent.com/pod-product-compliance
Lightning Source LLC
Chambersburg PA
CBHW022058170526
45157CB00004B/1392